World of science

LIGHT AND VISION

BAY BOOKS LONDON & SYDNEY

1980 Published by Bay Books
157–167 Bayswater Road, Rushcutters
Bay NSW 2011 Australia
© 1980 Bay Books
National Library of Australia
Card Number and ISBN 0 85835 270 2
Design: Sackville Design Group
Printed by Tien Wah Press, Singapore.

LIGHT

Light and the eye

Our eyes are sensitive to light, which gives us information about the shapes, colours, and movements of objects around us. Light is a form of *electromagnetic radiation* and we know it travels in the form of waves. The complete range of electromagnetic radiation is called the *electromagnetic spectrum* and our eyes are capable of seeing only part of the spectrum. We can see a large part of the wavelengths emitted by the sun, that is, white light but the sun also emits other waves which we cannot see.

Infra-red is a wavelength emitted by the sun which cannot be seen, though we can feel it in our bodies as warmth or heat. Ultraviolet is another form of light we cannot see, but we know about it because it tans our skin in summer.

How light behaves

Light moves in straight lines from its source, but it can be bent and scattered by objects placed in its path. We see rays of sunlight streaming through a window on a sunny day because some of the light is scattered by dust particles in the air. We can only 'see' a ray of light when it strikes the eye directly. Then it forms an image of the object from which it has come, either the light source itself, or something from which it has been reflected, such as a motor-car. *Non-luminous* objects are ones which are only visible when they reflect light from a light source. In a totally black room, you would not be able to see a desk,

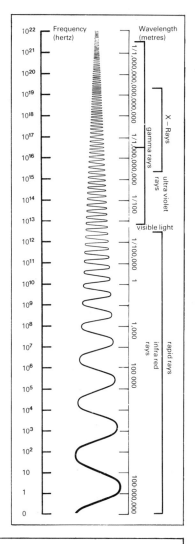

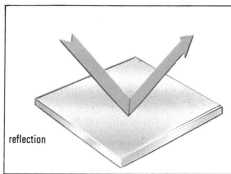

reflection

Left: Reflection of light. When light strikes an opaque, smooth surface, it is reflected back again at the same angle.
Right: Refraction of light. When light is passed from a less dense medium into a more dense medium and then back again, the rays are bent or refracted.
Above right: The electromagnetic spectrum.

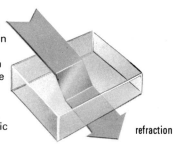

refraction

but you would be able to see the hands of a luminous clock. If the totally black room had no dust particles floating around it, you would not be able to see the *beam* of a flashlight, but only the light source itself and any object that reflects the light.

Speed of light

Even though an electric light appears to glow immediately it is switched on, a small but definite time lag occurs between the light coming on and the electromagnetic radiation entering our eyes. In a room, this time lag is too short to be noticeable, but for distant objects like stars, the lag is hundreds of thousands of years. Even light from the moon, which is relatively close to earth, experiences a time lag of one second. The speed of light, measured in a vacuum, is 299,792·5 km/sec (approximately 186,281 miles/sec). The speed is slower in denser media, such as air, water and glass.

COLOUR

The wavelength of light is related to its colour; red light, for example, has a longer wavelength than blue light. The light from the sun which we call white light is actually made up of different colours which combine to make white light. These colours are: violet, indigo, blue, green, yellow, orange and red.

The spectrum

By using a *prism* made of glass or plastic, it is possible to see the colours that make up sunlight. The colours separated in this way are called a *spectrum*. Another way to see the spectrum of sunlight is to look at a rainbow. In a rainbow you can see the colours quite clearly. A rainbow is caused by sunlight passing through raindrops. The light is bent, and because some wavelengths bend more than others, the colours are separated. The violet rays are bent the most and the red rays least.

Colour is not something which can be weighed and

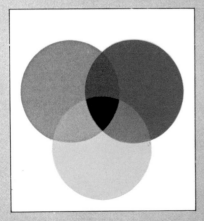

Colour by absorption (or subtraction) of light. Mixing cyan, magenta and yellow pigments (paints) together, yields a black effect.

measured in a laboratory, but is a sensation we experience when certain light waves affect the *retina* of our eyes. The prism experiment shows how white light is made up of a combination of wavelengths of different coloured light. To make colours it would seem that we would need paints or dyes of every possible colour and shade to get exactly what we want, but in fact any colour can be made by combining various proportions of the three basic colours. These are also called the *primary colours.*

Primary and secondary colours

The three primary colours in light are red, green and blue. White light can be made by mixing red, green and blue light. Yellow can be made by mixing green and red light.

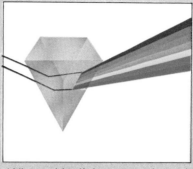

When white light passes through a prism in a laboratory experiment, you can see it is made up of different colours.

When mixed together, the primary colours – red, blue and green light – make up white light.

The process of making colours by mixing primary colours of light is called *addition,* because one colour is added to another. The colour in a colour TV is made in this way.

Mixing colours

This can easily be demonstrated by projecting a green beam and a red beam together on to a screen.

Colours made by combining two primary colours are called secondary colours. They are yellow (red & green), cyan (blue & green) and magenta (blue & red). When the primary colours are mixed in different proportions, any colour at all can be produced. Black is an absence of colour.

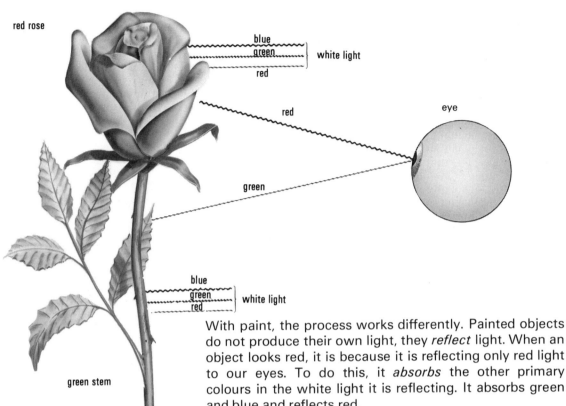

The flower of a rose is red because it absorbs all colours except for red which is reflected. The leaves and stem are green as they absorb the other primary colours and reflect only the green light.

With paint, the process works differently. Painted objects do not produce their own light, they *reflect* light. When an object looks red, it is because it is reflecting only red light to our eyes. To do this, it *absorbs* the other primary colours in the white light it is reflecting. It absorbs green and blue and reflects red.

A yellow object is absorbing only blue light but it is reflecting the red and green light which, as we know, make yellow. So any object we see that is reflecting light is actually absorbing some wavelengths first. The wavelengths that are reflected are what gives it the colour we experience.

The three primary colours used in paints and pigments or colouring substances are not the same as those of light. This is because light colours are formed by the process of *addition* and the colours of paint are formed by the process of *absorption* (or subtraction.) The three *primary pigments* or paint colours are actually the three secondary colours for light: **yellow (red & green), cyan (blue & green), magenta (red & blue).** Mixing yellow and cyan gives green, because yellow subtracts blue and cyan subtracts red. All that is left of the three primary colours to be reflected is green, and that is what we see.

Since paints absorb some of the wavelengths that fall on them, we can't make white by mixing paints. White is made up of all the colours. A special paint has to be used to produce white and in printing the white paper gives the necessary white effect. If all three primary colours are mixed together in equal amounts, then virtually all the wavelengths will be absorbed, leaving a dark brown, almost black colour. This is what happens with water paints when you mix too many of them together. The more varied the mixture the more wavelengths are absorbed and the darker the mixture becomes.

OPTICS

Optics is the study of light. It is concerned with the nature of light and the way it behaves in optical instruments. Light is a form of energy and so an object may only produce light when there is energy present. A red-hot piece of metal receives energy in the form of heat and converts some of it into red light. A TV screen receives electrical energy and produces light and a firefly or glow-worm converts the chemical energy of its body into light. The sun produces light from nuclear energy.

Reflection

When light strikes an opaque surface, like a table-top, some of it is absorbed and some of it is reflected. The light will be reflected at different angles depending on the type of surface. On a smooth surface, the light rays will be reflected back at the same angle and the surface therefore

In a car's driving mirror, like all mirrors, the reflection is reversed from left to right.

appears shiny. On a rough surface, the rays are reflected back at many different angles and so the surface looks dull.

We see objects like furniture because of the reflected rays and the direction from which the rays enter our eyes tells us where the object is. When we see things in a mirror, an image of the object is formed in the mirror though it is reversed from right to left.

Lenses

When light passes from one medium to another such as from air to water it is bent and this bending is called *refraction*. A common example is when a straw in a glass of water or soft drink appears to be bent. The straw does not bend at all, but the light is bent by the water and then reflected to reach our eyes. This is why it is so hard to spear a fish from above the surface of the water. Because of the refraction of light, the fish is not actually where you think it is and if you throw the spear straight in you will miss.

Another example of refraction occurs at sunset. When the sun has actually gone down below the horizon its rays are bent by the earth's atmosphere so that for a while it appears the sun is still shining.

When the sun sets its rays of light are bent or refracted by the atmosphere. So we can still see the sun for a little while after it has actually gone below the distant horizon.

seeing the sun after the sunset

setting sun

rays must bend in air

light refracted by atmosphere

atmosphere

earth

observer

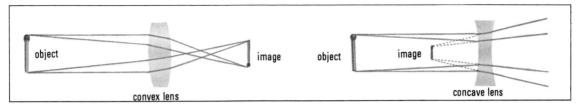

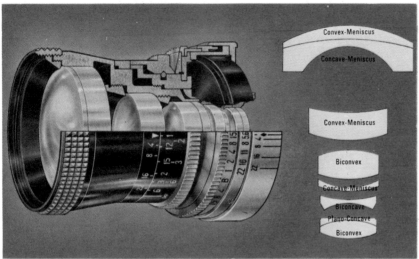

Refraction

Light can be bent by means of a lens. As it passes through the lens, the light is refracted and this effect can be used to produce visible images as in cameras, microscopes and telescopes. Glass is a common material for the manufacture of lenses because of its strength and clarity, but plastic is also used. The face of a lens can be made *convex*, that is bowed outwards, or *concave*, scooped inwards. In simple lenses, both faces are the same but in complex lenses one face may be convex and the other concave.

Positive and negative lenses

A convex lens causes light rays to *converge*, or come together, and is called a *positive* lens. A concave lens makes light *diverge*, or spread, and is known as a *negative* lens. A positive lens focuses light from a distant source into a visible image which appears on the opposite side of the lens to the object. For example, when a magnifying glass is used to focus the sun's rays to set fire to a piece

Top: These two lenses have opposite effects. The convex lens causes the light rays to converge. The concave lens makes them diverge. We see a real image produced by the convex lens, but the concave lens shows the object as a virtual image.

Above: This camera lens is made up of several different 'glasses' (components). Each 'glass' is shaped differently, and together they produce a bright and sharp image.

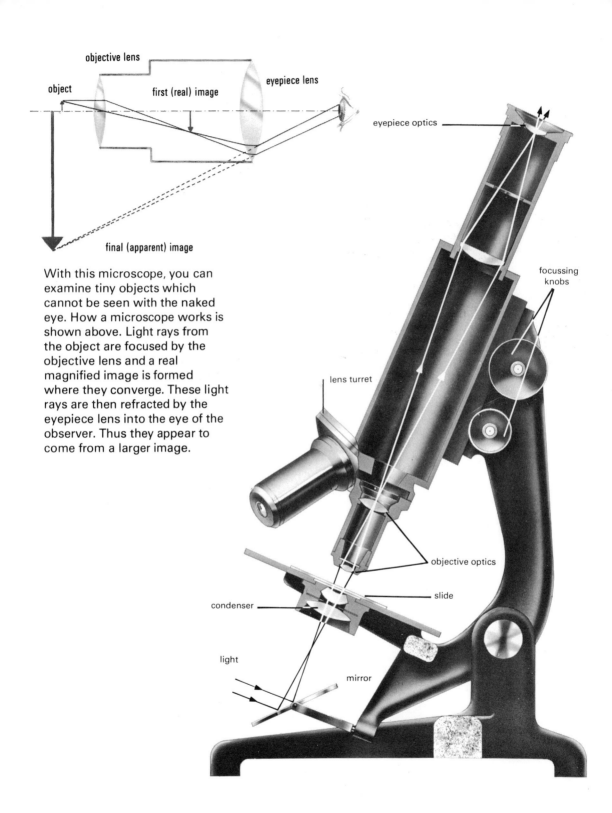

With this microscope, you can examine tiny objects which cannot be seen with the naked eye. How a microscope works is shown above. Light rays from the object are focused by the objective lens and a real magnified image is formed where they converge. These light rays are then refracted by the eyepiece lens into the eye of the observer. Thus they appear to come from a larger image.

of paper, it is making a small image of the sun. Making an image in this way is called *projection*.

A negative lens spreads the light as if it were diverging from a point on the same side of the lens as the object. The image that this makes is called a *virtual image*. It is not a real image and it cannot be projected.

The amount that a lens bends the light depends on the amount of curve on the faces of the lens. The distance between the lens and the image it produces is called the *focal length*. The shorter the focal length, the smaller the image. The greater the curvature of the faces of the lens, the shorter its focal length will be.

Optical instruments make use of lenses and mirrors to produce sharp images of objects that are either magnified or reduced. Generally, a camera can be attached to the optical instrument to take a photograph of the object.

The microscope

Microscopes are used to view and examine small objects which can't be seen with the naked eye. In the optical microscope light reflected from the object to be examined is beamed through a lens and this produces a magnified image. An ordinary magnifier is really a simple microscope which magnifies in this way. But in a *compound microscope*, another lens, held in the eyepiece, further magnifies the image so that high, or very high, magnification is obtained. Many microscopes have a *turret* at the bottom end which contains two or three different lenses for different magnifications. They can be swung into position as required. On some microscopes, the eyepiece contains only one lens so that you have to close one eye and look through it with the other; others have twin eyepieces so that both eyes can be used for a more comfortable view. They are called *binocular* microscopes.

The telescope

The telescope is the basic instrument of astronomers, who study the stars and planets. It was first known about 1608 and was developed into a practical instrument by the great Italian astronomer, Galileo, shortly afterwards. A basic optical telescope consists of two simple convex lenses. One, the objective, is of long focal length, while the other the eyepiece, is of short focal length. Working together, they give a magnified picture of distant objects.

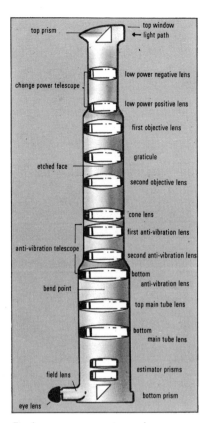

Periscopes are not newly invented – they were probably first used in the seventeenth century. This sophisticated, modern periscope is used in submarines to view the surface of the sea.

Field-glasses or binoculars are really twin telescopes of comparatively low power, in which prisms are used to bend the light rays so as to 'fold' the path of the light between the lenses and reduce the length of the instrument.

The telescopes used by astronomers are of two types, *refracting,* like Galileo's where the light is bent by the lenses to form enlarged images, and *reflecting,* where a curved mirror takes the place of the lenses in the refracting telescope.

Generally, the largest astronomical telescopes are reflectors, like the famous one at Mount Palomar, California, which has a reflector 200 inches or about 508

The large aperture diameter of this reflecting telescope means that it can probe deep into space. These powerful telescopes can study stars and planets so far away that their distances are measured in light years, not in kilometres.

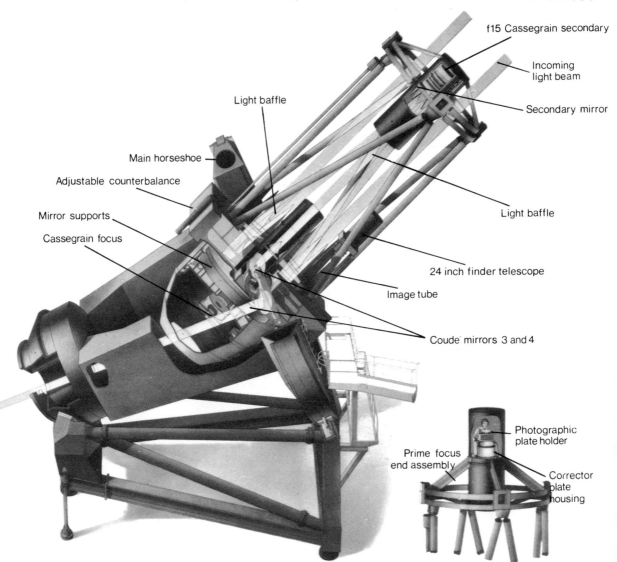

cm in diameter. There are few refracting telescopes of more than about 40 inches, or 100 cm in diameter. Large telescopes are used at Australia's Mount Stromlo observatory.

Radio telescopes, which are really radio receivers able to pick up and interpret the radio waves that are given out by distant stars and galaxies, are used today for research that is impossible with optical telescopes. A well-known radio telescope is the one not far from the town of Parkes, New South Wales.

Above: This giant radio telescope at Parkes in Australia, is also used as a radio receiver.
Below: This reflecting telescope was invented by Newton in 1668 to overcome blurred images.
Below left: Since Newton's first telescope, there have been many other optical developments.

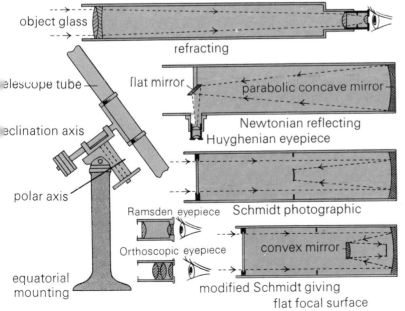

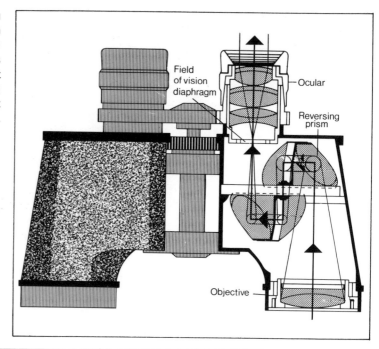

Cross-section of binoculars. You can see images of greater depth through binoculars than through a telescope. Each tube contains two prisms which bend the light four times as it passes through to give a close-up view of distant objects.

SIGHT AND THE EYE

Sight is a sense that is vital to our survival. Our eyes warn us of danger, lead us where we wish to go and tell us who our friends and enemies are. What we can see with the naked eye gives us a good understanding of our surroundings; what we can see with the aid of optical instruments such as telescopes and microscopes lets us examine the stars and the structure of atoms and molecules.

The eye

The eye is a round ball about 25 mm in diameter resting in a bony socket. A group of eye muscles allows us to turn our eyes in almost any direction. When you look at another person's eyes, the parts you see are: the 'white', the iris, and in the centre of the *iris*, the pupil. At the back of the eye is the *retina* which contains the light-sensitive nerve endings which send signals to the brain along the *optic nerve*.

The lens of the eye and the *cornea* in front of it enable us to focus light from a scene in front of the eye onto the retina at the back of the eyeball.

The shape of the lens is altered by a set of muscles called the *ciliary* muscles. When we look at close objects, the lens is altered by these muscles until it becomes rounded. When we look at distant objects, the lens becomes flattened. This change in the lens which allows us to see things clearly and sharply, whether they are near or far away, is called *accommodation*. The effect of accommodation is the same as the effect of focusing a camera lens on the main object of the picture you wish to take. Accommodation is a completely automatic or *reflex* action, so we do not have to think about moving our ciliary muscles when we look at near or far objects.

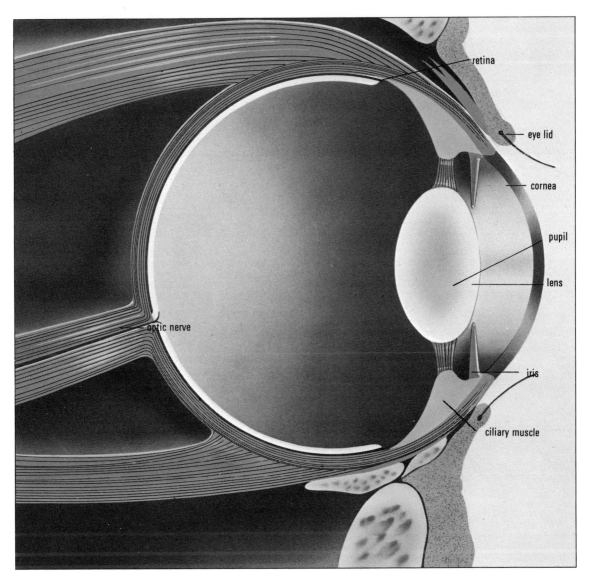

The human eye: behind the pupil and iris (the parts you can see) are the lens and the retina which contains light-sensitive nerve endings. Light enters the eye and is refracted by the cornea and lens, and finally focused on the retina.

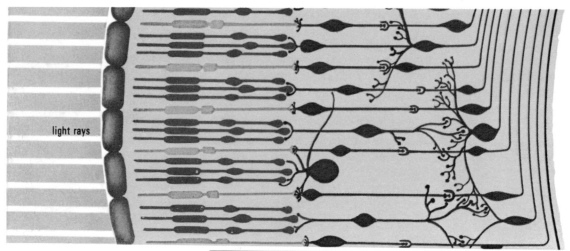

Above: The retina contains about 130 million light-sensitive nerve endings (rods and cones). The rods (shown in red) can distinguish only shades of grey while the less numerous cones (shown in blue) can distinguish colour.

Below: How the eye focuses. Light rays from a faraway object are focused on the retina. The lens is stretched thin by the taut fibres. When light rays from a near object enter the eye, the lens thickens to make the light rays converge.

The retina

When we look at an object, light from the object is focused on the retina which forms an image or picture of what we are looking at. The retina of each eye has about 130 million light-sensitive nerve endings. Most of these are comparatively long and are called *rods*. There are about 7 million shorter ones called *cones*. Rods are spread over the retina, whereas cones are almost all in the centre of the retina. It is believed that rods cannot distinguish between colours, but they are very sensitive to light generally and help us to see in poor light. Cones are less sensitive to light but can distinguish colours. Because the cones are less sensitive to light, it is difficult to recognise colours in dim light. There is a spot right at the back of the retina where the optic nerve is attached and there are no rods or cones. This is called the blind spot.

How we see

The way in which light is turned into nerve signals is not completely understood but is believed to involve a pigment called *rhodopsin*, or *visual purple*. This is located

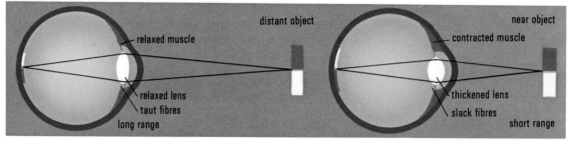

in the rods, and there are three similar pigments in the cones. Light causes molecules of rhodopsin to split in two and this causes a nerve impulse which is passed to the brain. Later, the rhodopsin molecules reform, ready to receive more light.

Nerve signals activated by the rods and cones are transmitted to the optic nerve which carries them to the *occipital lobes* at the back of the brain. On the way, the optic nerve passes through a nerve junction called the *optic chiasma,* where the views received by our two eyes are combined. By comparing the two slightly different images, the brain is able to work out depth and distance.

Colour-blindness

Colour-blindness is inability to see colours correctly. Usually a colour-blind person can see some colours without difficulty but confuses others. In a common type of colour-blindness there is difficulty in seeing red, either by not being sensitive to red at all, or confusing it with green. Colour-blind people are sometimes aware that they are seeing things differently from the rest of us. When they look at traffic lights, for example, they may not see the same colours as red, amber, green, but they must tell by the position of the lights that one means 'stop' and another means 'go'.

Colour-blindness can develop from infection or damage to the eye, but is usually inherited. A man can inherit

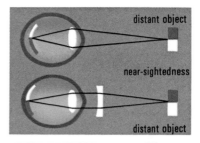
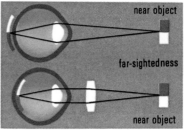

Above: Lenses are used to correct near (short) sightedness and far (long) sightedness. A concave lens corrects near-sighted vision by diverging the light rays so that an image is formed on the back of the retina. A convex lens corrects far-sightedness by converging rays.

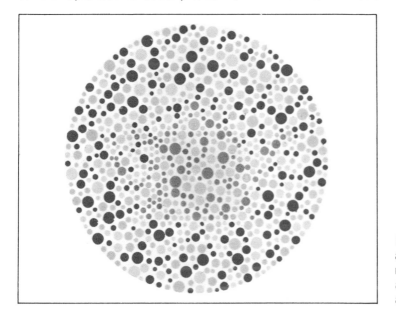

Left: Use this chart to test if you are colour-blind. If you have normal colour vision, you will see a cup. If you see a teapot, you are colour-blind.

colour-blindness from either parent but a woman must inherit it from both parents. As a result, more men than women are colour-blind.

Special test cards are used to determine the extent to which a person is colour-blind. The test cards generally have a pattern of grey and coloured spots, designed so that a colour-blind person cannot see a particular pattern in the spots that a person with perfect eyesight can see.

Short-sightedness

Apart from colour-blindness, a number of other defects may affect perfect vision. Some are diseases which can result in total loss of sight. Common defects which most of us know about are short-sightedness and long-sightedness.

Other names for short-sightedness are near sight or *myopia*. The eye can focus on very close objects but not so well on distant ones. A short-sighted person who wears glasses will often take the glasses off to read small print in a book or newspaper. In short-sightedness, the eyeball may be too long, or the lens or cornea may bulge too much. This causes the image to be formed slightly in front of the retina and it is blurred. This defect is fairly common in children and may be inherited. It can be corrected by wearing glasses or contact lenses with concave lenses.

Long-sightedness

Other names for long-sightedness are far sight or *hypermetropia*. This is exactly the opposite to short-sightedness and with this defect the eye can see distant objects clearly but has difficulty with close objects. People with this defect will put on their glasses to read. Long-sightedness results from too short an eyeball, or too flat a cornea or lens. In this case, the image comes to a focus behind the retina. This can be corrected by wearing glasses or contact lenses with convex lenses.

In many people, as they get older, the lens of the eye becomes stiff and less elastic. When this happens, the lens tends to stay more or less in the same shape and does not accommodate easily. The shape it stays in is better for long-sight than short-sight and so many older people have reading glasses. When the lens becomes stiffened in the middle-distance range, it is necessary to

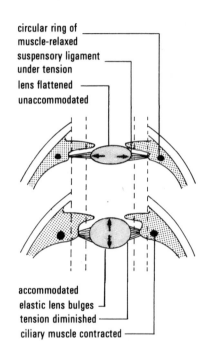

circular ring of muscle-relaxed
suspensory ligament under tension
lens flattened unaccommodated

accommodated
elastic lens bulges
tension diminished
ciliary muscle contracted

In an unaccommodated eye, a lens is flattened due to tension. For accommodation to occur, the ring of ciliary muscle must contract, thereby reducing tension and causing the lens to bulge. Then close and distant objects can be clearly seen.

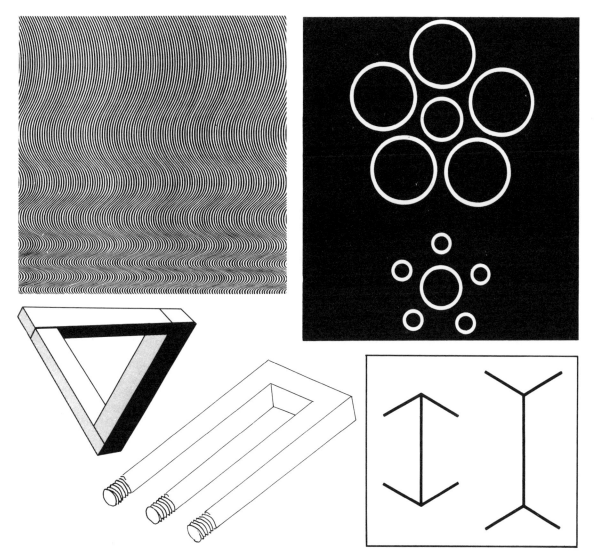

have two pairs of glasses, one for reading and one for long-distance. Bifocals combine two different lenses in the one frame.

Optical illusions

An *optical illusion* occurs when the brain makes a wrong interpretation of what the eye actually sees. This is not the same as an *hallucination,* which is something a person thinks he sees, but which does not actually exist.

Optical illusions are quite normal, and are often seen by many people, though the effect is stronger for some people than for others. Experiments have shown that

Top left: Scan these black, wavy lines and they seem to move.
Top right: The circle surrounded by smaller circles looks larger than the one ringed by bigger ones. But look closely and they are both the same size.
Above: The vertical line on the left appears shorter than the one on the right, but they are, in fact, both the same length
Above far left and left: The fork and triangle play tricks on your eyes – it is called reversible perspective.

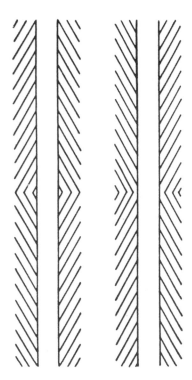

Another optical illusion, or is it? These lines appear to spread apart (right) and come together (left) but they are both parallel.

some optical illusions occur even in animals such as pigeons and fish.

The best-known optical illusions involve shape and size, although we can sometimes be misled by movement, position and colour. In what is known as the *Muller-Lyer arrow illusion* the two vertical lines are actually the same length even though the one with the arrowheads pointing outwards appears shorter than the one with the arrowheads pointing inwards.

In the circle illusion, the circle surrounded by smaller circles looks bigger than the one surrounded by larger ones. If you measure the circles you will find they are exactly the same diameter, but the eye is deceived by the different sizes of the surrounding circles.

Illusions of shape occur strongly when lines or circles are drawn against or close to a background with a strong pattern. In bent line illusions, the horizontal lines appear bent though, in fact, they are all straight. A circle or square can be made to appear distorted or out of shape by placing it on a patterned background. This is often noticeable in the patterns in materials for shirts and dresses. Horizontal lines make a person look plumper, and vertical lines or stripes have the opposite effect. The same thing applies to furnishing fabrics. Notice how horizontal stripes will make a lounge suite look longer or vertical stripes on a lamp shade will make it look taller.

In many optical illusions, the brain mistakes part of what it sees for something similar. In a drawing of a cube, the brain can see either of two faces of the cube as being the one nearest to the eye. Staring at the drawing, first one face and then the other will 'jump out'. Experiments of this type indicate that our past experience in looking at things has an effect on what we see. For example, size illusions apparently result from a false interpretation of depth and the brain's unconscious calculation of what it thinks is the 'true' size of the object.

THE CAMERA

Opposite: The modern camera is a highly developed mechanism. This cross-section view through the camera shows how it works.

The human eye and the camera have many similarities. All cameras, from the simplest to the most complicated, work on the same basic principle. Light from a scene enters a closed box through a hole or lens and throws an image or picture of the scene on to a light-sensitive film at the back of the box. This is called *exposing* the film. The exposed film is removed from the camera in the dark, and is

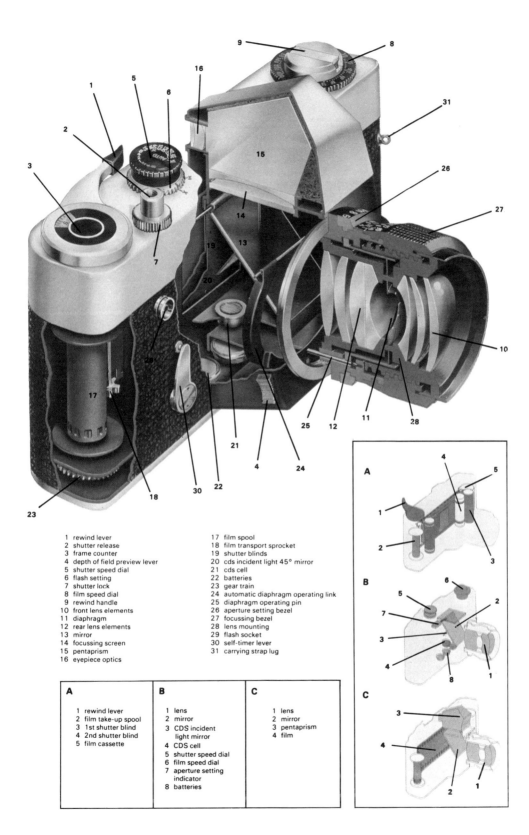

1 rewind lever
2 shutter release
3 frame counter
4 depth of field preview lever
5 shutter speed dial
6 flash setting
7 shutter lock
8 film speed dial
9 rewind handle
10 front lens elements
11 diaphragm
12 rear lens elements
13 mirror
14 focussing screen
15 pentaprism
16 eyepiece optics

17 film spool
18 film transport sprocket
19 shutter blinds
20 cds incident light 45° mirror
21 cds cell
22 batteries
23 gear train
24 automatic diaphragm operating link
25 diaphragm operating pin
26 aperture setting bezel
27 focussing bezel
28 lens mounting
29 flash socket
30 self-timer lever
31 carrying strap lug

A
1 rewind lever
2 film take-up spool
3 1st shutter blind
4 2nd shutter blind
5 film cassette

B
1 lens
2 mirror
3 CDS incident light mirror
4 CDS cell
5 shutter speed dial
6 film speed dial
7 aperture setting indicator
8 batteries

C
1 lens
2 mirror
3 pentaprism
4 film

19

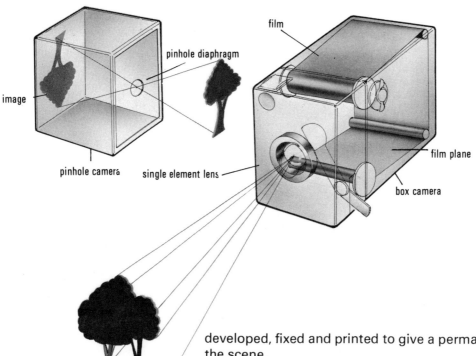

developed, fixed and printed to give a permanent record of the scene.

The pinhole camera

The simplest camera is a pinhole camera which consists of a box with a small hole in one of its sides. To produce a sharp image, the hole must be very small and this restricts the amount of light entering the camera. Quite a long time may be necessary to let enough light through to affect the film and this causes problems because if the subject moves the picture will be blurred. It is impossible to photograph anything like a moving car or a galloping horse with a pinhole camera.

If the hole is made bigger to let more light through and so reduce the time necessary for exposure, the image will be fuzzy. The light cannot be focused with a large hole

In the pinhole camera (above left), the tiny hole acts in the same way as a lens to produce a sharp image. The box camera (above right) is one of the simplest cameras. The single element lens focuses the light onto the film to produce a sharp image.
Below: A camera's shutter opens and closes to allow light to enter through the aperture.
Light rays are focused by the lens to form an image on film.

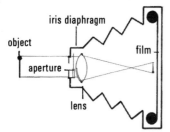

Cameras with lenses

A *lens* can be used to focus the light onto the film to produce a bright, clear image. The hole behind the lens is called the *aperture* and on many cameras the size of the hole, or aperture, can be altered. The length of time that light is allowed to enter the camera is called the *exposure* and is controlled by a *shutter*. In its normal position the shutter is closed and prevents light entering the camera. When the button is pressed, the shutter flies open for a pre-determined length of time, depending on the light conditions in which the photograph is being taken. This can be as long as one second or as short as 1/1000 second or even shorter. On a dull day you need a longer exposure than on a sunny day.

Both the diaphragm and the shutter need to be adjusted according to the amount of light that is available for taking a photograph. At midday in summer there will probably be plenty of light. On a winter afternoon there may not. In a living room at night the light may be quite good for the eye, but not enough for the camera.

The shutter

The shutters on most cameras can be adjusted to different *shutter speeds.* The shutter speed means the length of time the shutter is open. This might be several seconds (or even hours if you are photographing the night sky) or one-thousandth of a second or even less with special cameras. Most cameras have a shutter speed dial showing speeds from one second to, for example, one-thousandth of a second. The dial is set to the speed the photographer wants. Of course, the 'faster' the shutter speed the shorter the time the shutter is open and the smaller the amount of light let in. Shutter speeds are arranged so that each setting will let in half the amount of light let in by the one below it and twice the amount of the one above it. There is usually also a time-exposure setting so that the shutter can be left open for minutes or even hours in certain conditions.

The diaphragm

The diaphragm controls the size of the aperture in the same way as the iris of the eye: if you look at a cat's eyes when it comes in out of the darkness you will see that the irises have contracted to make the pupils bigger. After a few moments in a bright light the irises expand and cause the pupils to become much smaller. The aperture of the camera must also be larger in dim light and smaller in bright light.

The diaphragm is usually a ring of overlapping metal leaves which can be adjusted. The control settings for the diaphragm are referred to as *f-stops* and going from one f-

The diaphragm controls the size of the aperture as shown.
Light passes through the lens on to the film. This lens can be moved in or out for focusing. Use a small aperture on a bright day, a large one on a dull day.

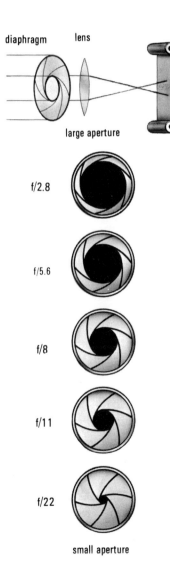

stop to the next reduces the amount of light by one half. The common setting are f/2.8, f/4, f/5.6, f/8, f/11, f/16, and f/22. Going from f/4 to f/5·6 admits half as much light.

Focusing

In the less expensive types of cameras, like the box type, the position of the lens is not adjustable. This means that a sharp image is focused on the film only when the subject being photographed is beyond about 2 or 3 m. These cameras are quite suitable for most home photographers, especially to record places we visit. But for close-ups the camera lens must be adjustable so that it can be moved outwards and away from the film. On most cameras this is done by means of a screw thread.

The lens

Similar to the lens of the human eye, a camera lens focuses the light. It can be made of glass or plastic. But whereas the eye can alter the shape of its lens from moment to moment by means of the ciliary muscles to allow for varying light conditions and distances, the shape of the camera lens is fixed and this causes certain types of *distortion.* These distortions are known as *spherical aberration* and *chromatic aberration.* To overcome such problems, a good-quality lens is made up of several lenses

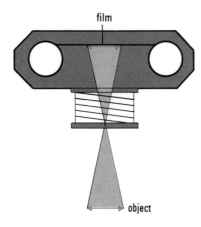
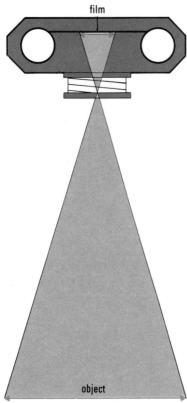

Above: The position of the lens can be adjusted to improve focusing.

Right: Both spherical and chromatic aberration are types of distortion caused by the fixed shape of a camera lens. Different lenses can be combined to overcome these types of distortions.

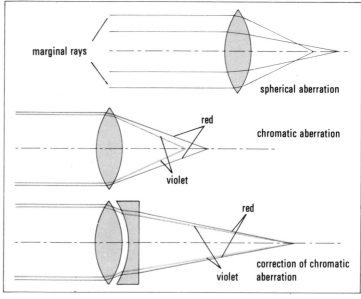

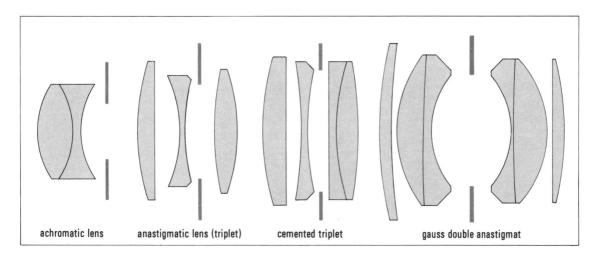

achromatic lens   anastigmatic lens (triplet)   cemented triplet   gauss double anastigmat

joined together. In many cameras, the lens can be removed and replaced by a different kind for special conditions. A common example is the telephoto lens for photographing distant objects; like taking a photo through a telescope, it often gives a much clearer image than we can get with our own eyes. Excellent lenses are now made for photographing distant objects and satellite photographs give us views of the earth we could not possibly see if we were in the satellite.

Above: These are all basic forms of photographic lenses. Note the very different shapes.

Below: This telephoto lens which revolutionized photography, is used for photographing distant objects. You can focus accurately with such a lens.

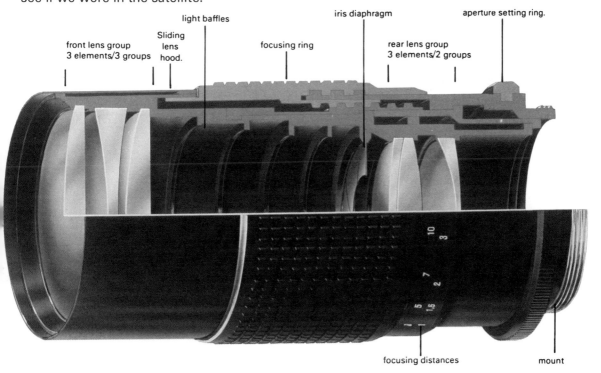

PHOTOGRAPHY

Taking good photographs is made easy by the modern camera. Virtually all you have to do is load the camera with film, make one or two adjustments for exposure and aperture, aim and 'shoot'.

Film

Photographic film is a strip of cellulose acetate plastic, sometimes with a protective paper backing. The plastic is coated with a substance that is sensitive to light, which is usually an *emulsion* of silver salts in gelatine. When exposed to light, these salts change chemically and the amount by which they change depends on the amount of light reaching them. We cannot see the change that takes place in the salts but when the film is *processed* – developed and printed – we have a visible picture.

The amount of light reaching the film is probably the most important part of good photography. Manufacturers make photographic film with varying degrees of sensitivity to light to let you choose film for different lighting conditions such as on the beach or the snow or indoors.

Above: Photography has come a long way since this early photograph was taken. Note the lack of definition and sharpness and how the tones seem to be merging into each other

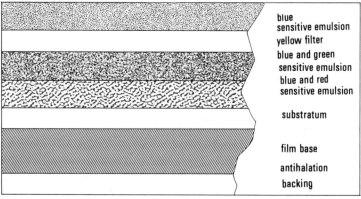

Right: This cross-section view through a colour film shows its varying degrees of sensitivity to coloured light.

Film sensitivity to light is referred to as its speed, so fast film is extremely sensitive to light and may be suitable for times when there is not much light, such as photographing birds in a forest. Fast film requires a shorter exposure than slow film. So fast film is suitable for action photography where, to prevent the picture being blurred, you want to use a fast shutter speed. Slow film needs a longer exposure time for the same light conditions.

Film speed

Film speeds are expressed by two common systems, one being the American Standards Association, or ASA, the other being the German DIN. The instructions that come with a roll of film usually include a table which shows both speeds. For example, 50 ASA is the same as 18 DIN (slow film), and 200 ASA is the same as 24 DIN (fairly fast).

Exposure

In taking photographs combinations of exposure and aperture are selected depending on the subject and the light conditions. To photograph a galloping horse, for example you may wish to use a fast exposure say 1/500 second, in order to freeze the horse in action. From the instructions with the roll of film, or by information from an

The shutter controls the time exposure — the amount of light entering the camera. This focal plane shutter is fitted to most interchangeable lens viewfinder cameras. The shutter's positioning within the camera is shown at the bottom.

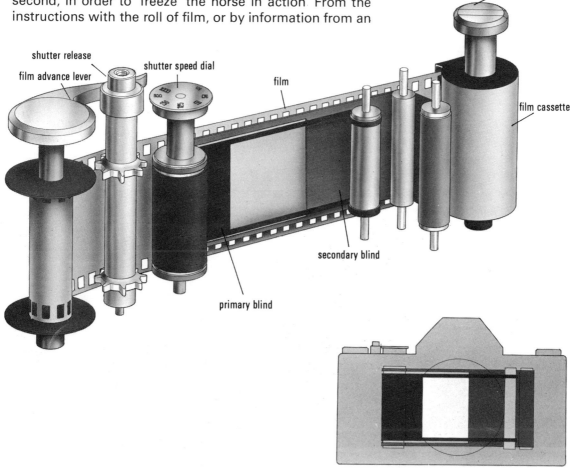

exposure meter, you may find you need an aperture of around f/4 in order to let enough light through.

This combination will work, provided you don't mind having the background blurred through being out of focus. This happens because the 'depth of field' is shallow at large aperture settings. 'Depth of field' is the range of distance through which objects are clear and sharply defined. It means that when the camera is focused on the horse, objects both closer to you and further away will be out of focus. This may be overcome by reducing the aperture to, say f/16, but then in order to let enough light in you may need to slow the exposure to about 1/30 second. At this slow speed the horse will move some distance while the shutter is open and so appear blurred on the film.

At times like this, choosing a different film could work to your advantage. In this case, a fast film would allow a shot of the horse with the exposure at 1/500 second and f/16 and the light reaching the film through the small aperture should be sufficient.

This spread of modern cameras and attachments shows you what a wide range there is from which to choose. Choice of camera will be conditioned by how much you can afford and how serious a photographer you are.

Pentax

Hasselblad

Developing and printing

The light-sensitive coating of black-and-white film consists of a thin layer of tiny grains of silver bromide crystals. When these grains are exposed to light they break down and deposit dark grains of silver, the process being completed by the chemical action of the solution called the *developer*.

During development, the parts of the film most exposed to light deposit most silver and become dark, while those less exposed are not so dark. Where no light

The cameras are, from left to right: Pentax 35mm with changeable lens; Hasselblad with changeable lens; Mamiyaflex twin lens reflex with changeable lens; and the Polaroid SX70.

Mamiyaflex

Polaroid

Developing and printing your own films is not nearly as difficult as it looks. This series of illustrations shows how to develop a film in a standard developing tank. This page explains how to develop the film to obtain a negative. The process of making a print (positive) on special paper is pictured opposite.

at all reaches the film the grains remain unaffected and can be washed away, leaving clear film. When a film is developed, what was light in the original scene becomes dark and what was dark becomes light. So the developed film is called a *negative*.

To make a print, or a *positive,* the negative is placed over photographic paper which is coated with a light-sensitive emulsion. A light is turned on for a pre-determined length of time and passes through the negative according to the pattern of light and dark on it. What was light now becomes dark and what was dark becomes light. An enlarger, which works like a camera in reverse, is used to produce prints bigger than the negative.

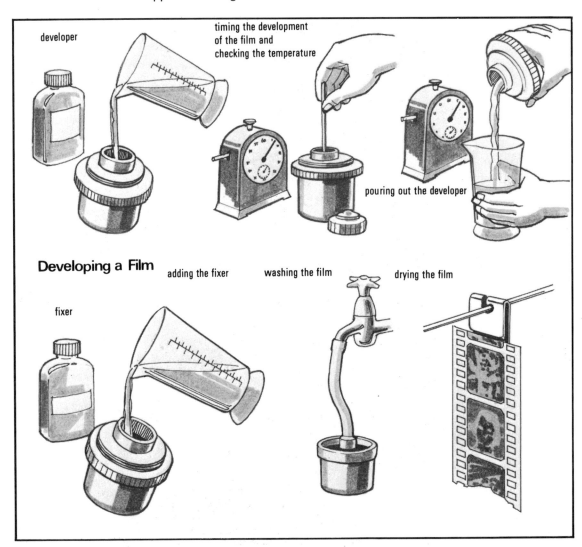

Developing a Film

developer — timing the development of the film and checking the temperature — pouring out the developer

fixer — adding the fixer — washing the film — drying the film

Making a Print

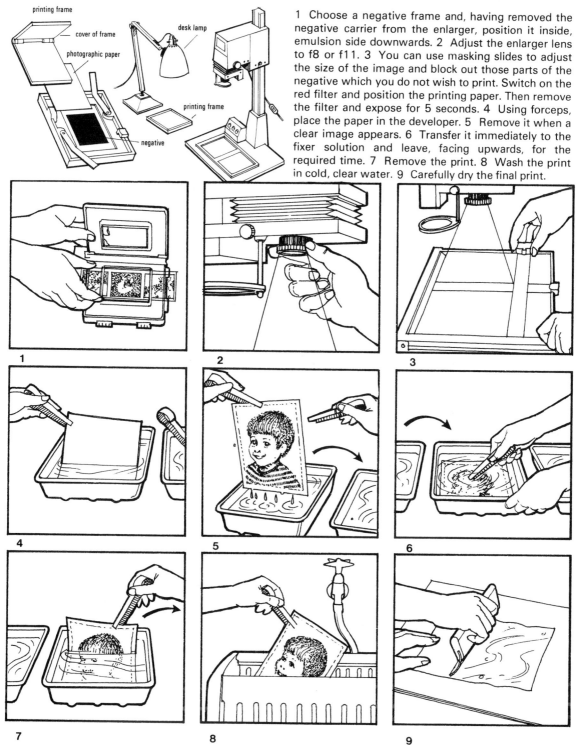

1 Choose a negative frame and, having removed the negative carrier from the enlarger, position it inside, emulsion side downwards. 2 Adjust the enlarger lens to f8 or f11. 3 You can use masking slides to adjust the size of the image and block out those parts of the negative which you do not wish to print. Switch on the red filter and position the printing paper. Then remove the filter and expose for 5 seconds. 4 Using forceps, place the paper in the developer. 5 Remove it when a clear image appears. 6 Transfer it immediately to the fixer solution and leave, facing upwards, for the required time. 7 Remove the print. 8 Wash the print in cold, clear water. 9 Carefully dry the final print.

Colour photography

There are two types of colour film. *Colour reversal* film produces colour positives such as slides and transparencies, which are viewed through a projector or viewer. Colour negative film is used to produce colour prints. Both films depend on the principle that any colour can be produced from different combinations of the three primary colours, blue, green and red.

A colour film is coated with three layers of light-sensitive emulsions; one sensitive to blue light, one to green and one to red. Light falls first on the upper emulsion layer which is sensitive to blue light. Beneath the blue-sensitive layer is a layer of yellow dye, which is included to absorb any blue light that may have got through. Then comes a layer sensitive to green light followed by a layer sensitive to red light. Each emulsion layer absorbs, or subtracts, a certain amount of the light passing through it. Hence, the process is called the subtractive process.

The Polaroid camera

Pioneered by an American inventor, the first black-and-white Polaroids were sold in 1948. These cameras are loaded with a double roll, a negative roll of film and a positive roll of special printing paper. In modern Polaroids and Kodak Instant cameras, after exposing the film to light through the lens, the photographic process is automatically controlled by miniature electronic circuits and very shortly after the shutter is pressed the print begins to emerge and develop, as a black-and-white or colour print, depending on the film.

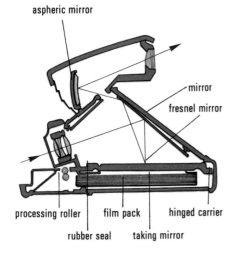

How a Polaroid camera works. When the film is exposed to light, two rollers break the chemical pods in the film, and a series of chemical reactions occur. Then a positive photograph forms.

Motion pictures

'Movies' are actually a series of still pictures projected onto a screen in quick succession to give the eye the impression of continuous motion. This happens because of what is known as *persistence of vision;* the eye retains for a moment the image of what it has seen. As a result, when a series of pictures are projected at the proper speed, they follow one another fast enough for the eye to join them together as one continuous image. If you look at a strip of movie film you will notice that each frame is only

series of pictures made in rapid succession

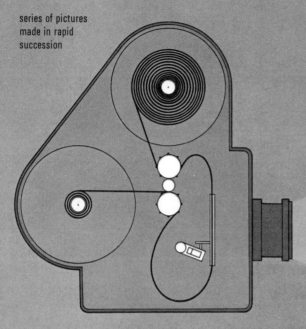

This is a motion picture camera. The film is passed in front of the aperture and the shutter opens and closes 24 times per sec. The film stops momentarily each time a picture is taken.

camera

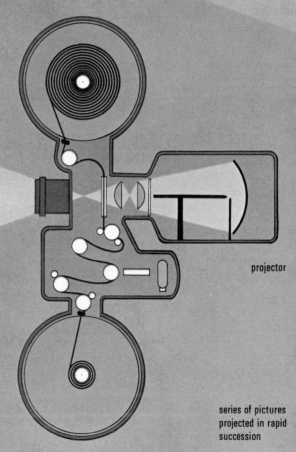

projector

A developed film is run through a projector which projects an image on to a screen, using a lens. You can view the film.

series of pictures projected in rapid succession

slightly different from the next one.

In the early days of movie photography the frames were shown at the speed of 16 frames every second. This is just too slow to produce a continuous effect and the pictures appear to flicker. These early movies became known as 'flicks'. In modern films, 24 frames are projected every second and each frame contains two separate but identical shots. This produces continuous movement without any flickering.

TELEVISION, VIDEOTAPES AND LASERS

The television camera

Experiments with television began in 1875 in America but it was not until 1926 that the world's first successful transmission of moving pictures was made in Britain.

A picture is televised by changing it to a series of electrical signals which are carried by radio waves to the receiver. There, the signals are used to re-form the original picture on the picture tube. On a television camera, a series of tiny photoelectric cells produce electrical charges

This television cameraman can film any events anywhere thanks to outside broadcast vans.

Opposite: How a programme is relayed from the studio to the television in your home.

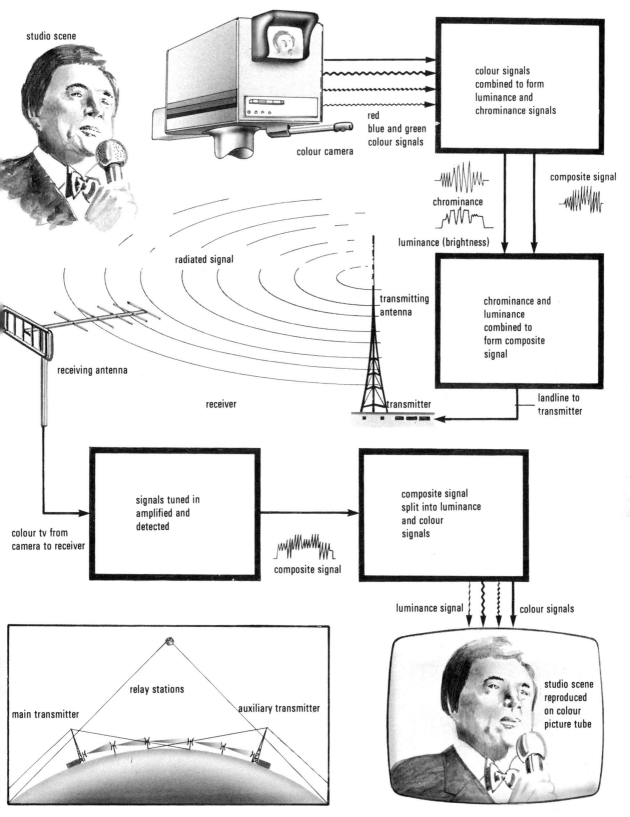

Above: This 'pop' group are getting ready for a live broadcast.

Below: In a black and white television screen, a combination of tones builds to create a picture.

according to the image. Brighter parts of the original image develop higher charges than the darker parts, so a pattern of varying charges is produced. This becomes an electrical copy of the image and it can be transmitted.

The television camera contains many thousands of photoelectric cells which have to be discharged, in the correct order, in a fraction of a second. To achieve this, a beam of electrons, produced within the camera, traces a path over the surface of the camera's screen. This is the process called *scanning,* and as the beam strikes each cell the charge of the cell is converted into an electrical impulse. This impulse is then ready for transmission. The scanning beam follows a particular order of movements and each time the camera screen has been completely scanned, the beam returns to its original position and repeats the process. The screen is completely scanned once every twenty-fifth of a second.

Moving objects are televised as a rapid sequence of still pictures or *frames.* Each frame shows the scene as it appears a fraction of a second after the sequence in the

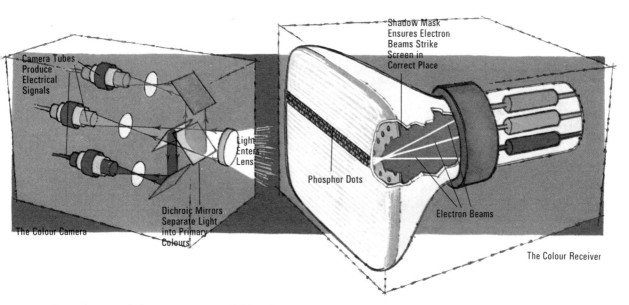

previous frame. It happens so quickly that our eyes see it as a continuous movement. The sound picked up by the microphone is transmitted at the same time but by a different carrier wave.

Unlike radio waves, television waves can be blocked by obstructions such as tall buildings and mountains. Television stations place their sending antennae on the tops of hills to try to overcome this problem and people living in valleys need tall receiving antennae to catch the signal. The problem can be overcome by using repeater antennae, erected so that they are 'in line of sight' from each other, or cables. With satellites, signals can be sent around the world.

The television receiver

The process of getting the television picture onto the screen in your home is the reverse of the way it was sent. The most important part of the TV set is the picture tube, or cathode ray tube. The wider end of the tube is the screen which is coated with a substance which emits light when it is struck by fast-moving electrons. The signals sent by the station are converted back to impulses which increase or decrease the intensity of the electron beam as it scans the back of the tube. The effect is to produce bright areas and dark areas on the screen. These areas form a replica or exact copy of the image picked up by the camera.

Above: Electrical signals of each primary colour are produced in the camera. Electron beams are motivated by the receiver and cause phosphor dots in the screen to glow red, blue and green, and to make up a colour picture.

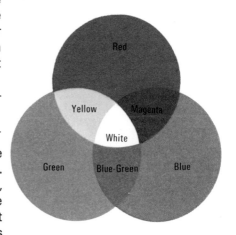

Above: Red, blue and green are the three primary colours. When mixed, they make other colours.

35

Colour television

In colour television, light is split into the three basic colours, red, green and blue. In your receiver, the television signal is separated into impulses which correspond to these colours and are made to vary the intensities of three separate electron beams, one for each colour, which scan the receiver's screen. The screen's surface is composed of a large number of sensitive dots arranged in groups of three. In each group, one dot glows red, one glows blue and the other glows green when they are struck by the electron beam. The colour produced by each group of dots depends on the intensities of the electron beams as they strike the group.

Colour televisions are now owned by many homes. They are housed in attractive cabinets.

Videotape recording

Television programmes can be recorded on *videotape,* a magnetic tape which records both sound and picture signals. The videotape recorder is similar in many ways to an ordinary sound tape recorder. For recording, the picture images are first changed into electrical signals. This can be done through the television camera and the electrical impulses fed into the recorder. Here, they are converted into magnetic variations and transferred to a moving tape with a magnetic iron-oxide coating. To play back, the magnetic signals are re-converted into electrical impulses which can be fed into a TV apparatus which reproduces the original scene on its screen.

There is now a wide range of equipment and luxury items which you can enjoy in your own home such as televisions, videotape recording equipment and cameras and cinecameras.

The tape used for videotape recorders is much wider than the tape on a sound tape recorder. Portable systems use either 12 or 25 mm tape, while the TV stations usually use 50 mm tape. On a typical videotape, the sound signal is recorded on one narrow track at the top, the video track occupies the central part, and a control signal is recorded along the bottom.

Lasers

The word laser stands for Light Amplification by Stimulated Emission of Radiation. The first laser was produced in 1960. A laser produces energy in the form of an intense beam of light. Light is made up of small parcels of energy called *photons.* When light comes from a lamp or from the sun, these photons are vibrating out of step with each other, like people walking down the street. But the photons can be brought into step, like a column of soldiers marching, and when this happens, the beam

Below: This powerful laser beam is being used to light a very ordinary cigarette.

Opposite: Laser beams can be used to 'burn-drill' small holes in a tough sheet of steel.

Laser beams can be attractive to look at as well as being functional. These laser beams are criss-crossing Oxford Street in London as part of the 1978 Christmas illuminations.

becomes very intense. This is the principle of the laser. Some lasers give a continuous beam of light, while others give a pulse of light lasting for a fraction of a second. Although the pulse lasts for a short time, its intensity is very great.

Lasers can drill holes in very hard substances and they can be used to weld and cut metals. Some lasers are used in surgery where they can be adjusted for very delicate operations, such as mending the retina of the eye, and others are being developed to carry communications in a similar way to radio and television.

INDEX
LIGHT 1-2
COLOUR 2-5
OPTICS 5-12
SIGHT AND THE EYE 12-18
THE CAMERA 18-23
PHOTOGRAPHY 24-32
TELEVISION, VIDEOTAPES AND LASERS 32-40

Page numbers in italics refer to a diagram on that page.
Bold type refers to a main heading or sub-heading.

A
Aberration (photography) 22, *22*
   chromatic 22, *22*
   spherical 22, *22*
Accommodation (eye) 13, *16*
Addition (light) 3, *3*
Antennae 35
   repeater 35
Aperture (camera) 20, *20*, 21, *21*, 25, *25*, 26
ASA (American Standards Association) 25
Astronomers 9
Atmosphere 6, *6*

B
Beam of light 2
Binocular microscope 9
Binoculars 10, *12*
Black 2, 3
Blind spot (eye) 14
Blue 2-3
Box camera *20*, 22
Broadcast, live *34*
Broadcast vans *32*

C
Camera 7, *7*, 9, **18**, 18-23, *19*, *26*
   aperture 20, *20*, 21, *21*
   box *20*, 22
   diaphragm *20*, **21**, *21*
   focusing 22
   f-stops *21*, 21-22
   lens 20, *20*, 21, **22**, *22*, 23, *23*
   motion 30, **31**
   pinhole **20**, *20*
   shutter 20, *20*, **21**
   shutter speed 21
   television **32**, 32-35, *33*, *37*
Cathode ray tube 35
Cellulose acetate plastic 24
Ciliary muscles 13, *13*, 16
Cinecameras 37
Colour **2**, 2-5, *2*, *4*, 14, *35*, 36
   absorption 2, *4*
   addition 3-5, *3*
   basic 3, 36
   black 2, 3-5
   blue 2-3
   cyan 2, 3-4
   green 2-4 *4*

magenta 2, 3-4
mixing 3
primary **3**, 3-5, *4*, 30, *35*, 36
red 2-5, *4*
secondary **3**, 3-5
yellow 2-4, *2*
Colour-blindness **15**, *15*, 16
Colour negative film 30
Colour photography **30**
Colour reversal film 30
Concave lens 7-9, *7*, 16
Cones (retina) 14-15, *14*
Contact lenses 16
Convex lens 7-9, *7*, 16
Cornea (eye) 12, *13*, 16
Cyan 2-3, *2*

D
Depth of field 26
Developer 27, *28*
Developing (film) **27**, 27-29, *28*
Diaphragm (camera) *20*, **21**, *21*
DIN 25
Distortions (photography) 22-23, *22*
   chromatic aberration 22, *22*
   spherical aberration 22, *22*

E
Electrical signals (colour) *35*
Electromagnetic radiation 1-2
Electromagnetic spectrum 1
Electrons 34-36
Emulsion 24
Energy 5
   chemical 5
   electrical 5
   heat 5
   light 5
   nuclear 5
Enlarger (photographic) 28, *29*
Exposure 24, **25**, *25*, 26
Exposure meter 26
Exposure time 20-21, 25
Eyepiece (microscope) *8*, 9
Eyes **1**, 2, **12**, 12-18, *13*, *14*, 16
   cornea 12-13, *13*, 16
   human *13*, 18
   iris 12, *13*, 21
   lens 12-13, *13*, 14, 16, *16*

muscles 12, *13*, 16
optic nerve 12, *13*, 14-15
pupil 12, *13*
retina 3, 12, **14**, *14*, 15, 40

F
Field-glasses 10
Film 18, **24**, *24*, 25, 27-28, 30, *31*
   black and white 27
   colour *24*, 30
   development **27**, 27-28, *28*
   fast 26
   light-sensitive 18, 24
   movie 30-32, *31*
   negative 28
   positive 28
   sensitivity 24, *24*
   speed 24, **25**
Firefly 5
Fixer 28
Flashlight 2
'Flicks' 32
Focal length 9
Focal plane shutter 25
Focusing
   camera 22, *22*
   eye 16, *16*, **22**
Frames (film) 34-35
F-stops 21-22, *21*

G
Galileo 9-10
Glass 7, *7*
Glasses (eye) 16
   bifocals 17
Glow-worm 5
Green 2-3

H
Hallucination 17
Hasselblad *26*
Heat 1
Hypermetropia 16

I
Images 7-9, *8*
   magnified *8*, 9
   real 7, *7*, *8*
   reduced 9

television 33-36, *34*
  virtual 7, *7*, 9
Indigo 2
Infra-red light 1
Iris (eye) 12, 13
Iron oxide 37

K
Kodak instant cameras 30

L
Lasers **38**, *38*, *39*, 40, *40*
  use 38-40
Lenses **6**, 7-11, *15*, 16
  camera **20**, *20*, 21, **22**, *22*, 23, *23*
  concave 7, *7*, 9, *15*
  contact 16
  convex 7, *7*, 9, *15*
  eye 12-13, 16
  eyepiece *8*
  negative **7**, 9
  objective *8*
  positive **7**, 9
  telephoto 23, *23*
Light **1**
  behaviour **1**
  blue 2-3, *3*
  colour **2**, *2*
  green 2-3, *3*
  indigo 2
  orange 2
  red 2-5, *3*
  reflected 4-6, *4*, **5**, *5*
  refraction 6, *6*, 7
  spectrum **2**, 3
  speed **2**
  violet 2
  wavelength 1-3
  waves 1, *3*
  white 1-5, *3*
  years 10
  yellow 2
Long-sightedness *15*, **16**
Luminous clock 2

M
Magenta 2-3, *2*
Magnifying glass 7-9
Mamiyaflex 27
Medium (air, glass) 6
Microscope 7, *8*, **9**
  binocular 9
  compound 9
  optical 9
Mirror *5*, 6, 9
Moon 2
Mount Palomar telescope 10-11
Mount Stromlo observatory 11
'Movies' 30-32
Myopia 16

N
Negative (film) 28
Newton's telescope *11*

Non-luminous object 1

O
Occipital lobes 15
Optical illusions **17**, *17*, 18, *18*
  bent line 18
  circle 18
  Muller-Lyer arrow 18
  size 18
Optical instruments 5, 9, 12
Optic chiasma 15
Optic nerve 12, *13*, 14-15
Optics **5**
Orange 2
Oxford Street, London *40*

P
Paints 2-5, *2*
Parkes, NSW 11, *11*
Pentax 26
Periscope *9*
Persistence of vision 30
Phosphor dots 35-36, *35*
Photoelectric cell 32, 34
Photography **24**, 24-32
  action 24-26
  aperture 25-26
  close-up 22
  colour **30**
  developing **27**, 27-29, *28*
  distortions **22**, *22*
  early *24*, 32
  film **24**, *24*, 26
  focusing **22**
  lenses **22**, 23, *23*
  motion **30**, 30-32, *30*, *31*
  printing **27**, 27-29, *29*
  satellite 23
Photons 38
Picture tube 32, *33*, 35
Pigments 2-5, *2*
Pinhole camera **20**, *20*
Polaroid (camera) 27, **30**, *30*
Positive (film) 28
Printing **27**, 27-29, *29*
Prisms 2-3, *3*, 10, *12*
Projection 9, 30, *31*
Pupil (eye) 12, *13*

R
Radio telescope 11, *11*
Radio waves 11, 32, *33*
Rainbow 2
Red light 2-3, *3*
Reflection **5**, 5-6, *5*
Reflex action (eye) 13
Refraction 6, *6*, **7**
Retina 3, 12, *13*, **14**, 15, 40
Rhodopsin 14-15
Rods (retina) 14-15, *14*
Rose, red *4*

S
Satellites 35

Scanning (television) 34
Seeing **14**, *14*, 15, **16**
  nerve impulses 14-15
Short-sightedness *15*, **16**
Shutter (camera) 20, *20*, **21**, *25*
Sight 12-18, **14**, *14*, 16
  colour-blindness **15**, *15*, 16
  long-sightedness *15*, **16**, 17
  short-sightedness *15*, **16**
Silver bromide 27
Silver salts 24
Slides 30
Spectrum **2**, 3
Stars 2
Subtractive process (film) 30
Sun 1, 5
  rays *6*, 7
Sunlight 1-2
Sunset 6, *6,*
Surfaces 5-6

T
Telephoto lens 23, *23*
Telescope 7, **9**, 9-11
  optical 9-11
  radio 11, *11*
  reflecting 10-11, *10*, *11*
  refracting 10-11
Television **32**, 32-38, *33*, **36**, *36*
  camera **32**, 32-35, *33*
  colour **35**, **36**, *36*
  picture tube 35
  receiver 32, *33*, *34*, **35**, 36
  signals *33*, 35, *35*
  studio *33*
  waves 35
Time lag 2
Transparencies 30
Turret (microscope) 9
TV (television) 3, 5

U
Ultraviolet light 1

V
Videotape **37**, *37*, 38
  portable 38
  recording **37**, *37*, 38
Violet 2
Visual purple 14-15

W
Warmth 1
White light 1-4, *3*

Y
Yellow 2-3, *2*